Macmillan/McGraw-Hill Science

EARTH'S oceans

AUTHORS

Mary Atwater
The University of Georgia

Prentice Baptiste
University of Houston

Lucy Daniel
Rutherford County Schools

Jay Hackett
University of Northern Colorado

Richard Moyer
University of Michigan, Dearborn

Carol Takemoto
Los Angeles Unified School District

Nancy Wilson
Sacramento Unified School District

Macmillan/McGraw-Hill School Publishing Company
New York Chicago Columbus

CONSULTANTS

Assessment:

Janice M. Camplin
Curriculum Coordinator, Elementary Science
Mentor, Western New York
Lake Shore Central Schools
Angola, NY

Mary Hamm
Associate Professor
Department of Elementary Education
San Francisco State University
San Francisco, CA

Cognitive Development:

Dr. Elisabeth Charron
Assistant Professor of Science Education
Montana State University
Bozeman, MT

Sue Teele
Director of Education Extension
University of California, Riverside
Riverside, CA

Cooperative Learning:

Harold Pratt
Executive Director of Curriculum
Jefferson County Public Schools
Golden, CO

Earth Science:

Thomas A. Davies
Research Scientist
The University of Texas
Austin, TX

David G. Futch
Associate Professor of Biology
San Diego State University
San Diego, CA

Dr. Shadia Rifai Habbal
Harvard-Smithsonian Center for Astrophysics
Cambridge, MA

Tom Murphree, Ph.D.
Global Systems Studies
Monterey, CA

Suzanne O'Connell
Assistant Professor
Wesleyan University
Middletown, CT

Environmental Education:

Cheryl Charles, Ph.D.
Executive Director
Project Wild
Boulder, CO

Gifted:

Sandra N. Kaplan
Associate Director, National/State Leadership
Training Institute on the Gifted/Talented
Ventura County Superintendent of Schools Office
Northridge, CA

Global Education:

M. Eugene Gilliom
Professor of Social Studies and Global Education
The Ohio State University
Columbus, OH

Merry M. Merryfield
Assistant Professor of Social Studies and Global Education
The Ohio State University
Columbus, OH

Intermediate Specialist

Sharon L. Strating
Missouri State Teacher of the Year
Northwest Missouri State University
Marysville, MO

Life Science:

Carl D. Barrentine
Associate Professor of Biology
California State University
Bakersfield, CA

V.L. Holland
Professor and Chair, Biological Sciences Department
California Polytechnic State University
San Luis Obispo, CA

Donald C. Lisowy
Education Specialist
New York, NY

Dan B. Walker
Associate Dean for Science Education and Professor of Biology
San Jose State University
San Jose, CA

Literature:

Dr. Donna E. Norton
Texas A&M University
College Station, TX

Tina Thoburn, Ed.D.
President
Thoburn Educational Enterprises, Inc.
Ligonier, PA

Macmillan/McGraw-Hill School Division
10 Union Square East
New York, New York 10003

Printed in the United States of America

ISBN 0-02-274265-4 / 4

3 4 5 6 7 8 9 VHJ 99 98 97 96 95 94 93

A coral reef

Mathematics:

Martin L. Johnson
Professor, Mathematics Education
University of Maryland at College Park
College Park, MD

Social Studies:

Mary A. McFarland
Instructional Coordinator of
Social Studies, K-12, and
Director of Staff Development
Parkway School District
St. Louis, MO

Physical Science:

Max Diem, Ph.D.
Professor of Chemistry
City University of New York, Hunter College
New York, NY

Gretchen M. Gillis
Geologist
Maxus Exploration Company
Dallas, TX

Wendell H. Potter
Associate Professor of Physics
Department of Physics
University of California, Davis
Davis, CA

Claudia K. Viehland
Educational Consultant, Chemist
Sigma Chemical Company
St. Louis, MO

Reading:

Jean Wallace Gillet
Reading Teacher
Charlottesville Public Schools
Charlottesville, VA

Charles Temple, Ph. D.
Associate Professor of Education
Hobart and William Smith Colleges
Geneva, NY

Safety:

Janice Sutkus
Program Manager: Education

National Safety Council
Chicago, IL

Science Technology and Society (STS):

William C. Kyle, Jr.
Director, School Mathematics and Science Center
Purdue University
West Lafayette, IN

Students Acquiring English:

Mrs. Bronwyn G. Frederick, M.A.
Bilingual Teacher
Pomona Unified School District
Pomona, CA

Misconceptions:

Dr. Charles W. Anderson
Michigan State University
East Lansing, MI

Dr. Edward L. Smith
Michigan State University
East Lansing, MI

Multicultural:

Bernard L. Charles
Senior Vice President
Quality Education for Minorities Network
Washington, DC

Cheryl Willis Hudson
Graphic Designer and Publishing Consultant
Part Owner and Publisher, Just Us Books, Inc.
Orange, NJ

Paul B. Janeczko
Poet
Hebron, MA

James R. Murphy
Math Teacher
La Guardia High School
New York, NY

Ramon L. Santiago
Professor of Education and Director of ESL
Lehman College, City University of New York
Bronx, NY

Clifford E. Trafzer
Professor and Chair, Ethnic Studies
University of California, Riverside
Riverside, CA

STUDENT ACTIVITY TESTERS

Jennifer Kildow
Brooke Straub
Cassie Zistl
Betsy McKeown
Seth McLaughlin
Max Berry
Wayne Henderson

FIELD TEST TEACHERS

Sharon Ervin
San Pablo Elementary School
Jacksonville, FL

Michelle Gallaway
Indianapolis Public School #44
Indianapolis, IN

Kathryn Gallman
#7 School
Rochester, NY

Karla McBride
#44 School
Rochester, NY

Diane Pease
Leopold Elementary
Madison, WI

Kathy Perez
Martin Luther King Elementary
Jacksonville, FL

Ralph Stamler
Thoreau School
Madison, WI

Joanne Stern
Hilltop Elementary School
Glen Burnie, MD

Janet Young
Indianapolis Public School #90
Indianapolis, IN

CONTRIBUTING WRITER

Molly Greenberg

Earth's Oceans

Lessons | Themes

Activities!

EXPLORE

TRY THIS

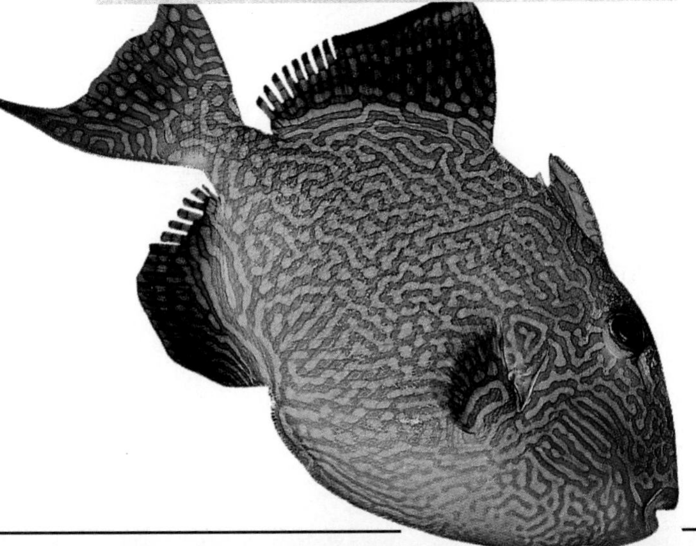

Features

 Links

Literature Link

Health Link

Social Studies Link

Music/Art Link

Language Arts Link

SCIENCE TECHNOLOGY AND Society

Focus on Environment

Focus on Technology

CAREERS

Departments

Earth's Oceans

"From sea to shining sea" describes any continent, because all land on Earth is surrounded by water. In fact, of all the planets in our solar system, Earth has the most water. Look at a globe. You can see that water covers over 70 percent of Earth's surface. Maybe we should call our planet Water or Ocean, rather than Earth.

Can you imagine being a part of a sailing expedition that would last three years? If you had been part of Ferdinand Magellan's crew in 1519, it would have taken you that long to sail around the world! That first successful voyage completely around the world proved that Earth is round and the oceans of the world are connected. These discoveries changed scientific thinking forever!

Indian Ocean

Minds On! Imagine you have a slice of apple in your hand. How does the thickness of the apple compare to the thickness of the peel? This relationship is similar to the one between the oceans and Earth. The apple represents Earth and the peel represents the oceans and continents. ●

Atlantic Ocean

TRY THIS

Activity!

Ocean Maze

What You Need

globe, *Activity Log* page 1

Locate the major oceans on a globe. Earth's largest oceans are the Atlantic, Pacific, Arctic, and Indian. Place a finger in the middle of one ocean. Then move your finger over the globe through all the other oceans and back again to the start. Do not lift your finger or cross land. Can you do it? Describe your route in your ***Activity Log.***

Pacific Ocean

You were able to move your finger around the globe without crossing land because the oceans are connected. Even though the oceans are connected, each ocean has its own characteristics. The Pacific Ocean is the largest—it's more than twice as large as the Atlantic. It also has some of the strongest storms, volcanoes, and earthquakes. Large masses of ice cover parts of the Arctic Ocean during most of the year. The temperatures of the water in the Arctic can be as cold as 28° F (about –2°C). With temperatures this low, why isn't the entire Arctic Ocean frozen?

Arctic Ocean

People have been asking questions about oceans for a long time. With so much water around, there have always been many areas to explore and a lot of questions to answer. What does ocean water taste like? What types of animals and plants live in the ocean? Why is it easier to float in the ocean than in a swimming pool? How can we clean up oil spills in the ocean? These are just a few of the questions you'll explore in this unit.

Scientific Method

Scientists use scientific methods to search for answers. A **scientific method** is not just a set of steps to follow like a recipe. Instead, it is an organized plan to solve science puzzles.

To help you in your experimenting, you can use the following five steps as your procedure.

1. ***State the problem***—What question are you trying to answer?
2. ***Write a hypothesis***—Give a guess or a possible answer to the question.
3. ***Design an experiment***—List the steps you will follow in order to test your hypothesis.
4. ***Record data***—Write down your observations during the experiment.
5. ***Draw a conclusion***—Based on your observations, report what happened. How did what happened compare with your hypothesis?

Using a scientific method to investigate a question or problem is necessary so that an experiment can be repeated. Repeating an experiment helps you be sure your answers are correct. Using a scientific method would also help someone else perform the same experiment. Do the Try This Activity on page 11 to practice using a scientific method.

Which Freezes Faster, Fresh Water or Salt Water?

When you use your knowledge and experience to solve a problem in a logical way, you are using a scientific method. Use a scientific method to find out more about ocean water.

Problem: Will fresh water and salt water freeze at the same temperature?

Hypothesis: Think about the problem and then add your hypothesis. Your hypothesis is your predicted answer and explanation for the question.

What You Need

2 16-oz. jars, permanent marking pen, thermometer, salt, small spoon, water, *Activity Log* page 2

Mark one jar with an F and fill it with water. Mark the second jar with an S and fill it with water. Stir 2 spoonfuls of salt into the second jar. Measure and record the temperature of the water in each jar. Place both jars in the freezer. After 24 hours, remove the jars from the freezer and record your observations. Measure and record the temperature of the salt water.

Conclusion: Describe what happened in your experiment. Was your hypothesis correct?

Now that you have solved the problem, you may think that you are finished with the activity, but this is just the beginning. After you have completed an experiment, you may have other questions to explore. Do you know why the fresh water froze but the salt water didn't? You might want to observe and compare the water samples to see if you can figure out why this occurs. This would be another problem. What is your hypothesis to the new question? What would your next step be?

Literature Link

Science in Literature

Down Under Down Under Diving Adventures on the Great Barrier Reef
by Ann McGovern.
New York: Macmillan, 1989.
See the shapes, sizes, and colors of ocean life that a 12-year-old sees while scuba diving in the waters near Australia's Great Barrier Reef. There are unusual unicorn fish, poisonous sea snakes, huge turtles, colorful clams, and hungry sharks. You'll learn about scuba diving and the treasures of ocean life. The ocean provided once-in-a-lifetime experiences for the divers. In return, the divers treated the Great Barrier Reef and its inhabitants with care so as not to disrupt the environment.

Other Good Books To Read

My Sister, My Science Report
by Margaret Bechard.
New York: Viking, 1990.
Even though Tess couldn't work on her science project with her best friend, she found out that science and her lab partner can be a lot of fun. What steps of a scientific method can you find in this book?

Life in a Tidal Pool
by Alvin and Virginia Silverstein.
Boston: Little, Brown and Company, 1990.
Find out what life is like at the edge of the ocean. Sea slugs, crabs, starfish, algae, sea lettuce, and jellyfish are some of the creatures that live in the quiet tidal pools. How do these plants and animals survive the waves, the weather, and each other?

Creatures of the Sea by John Christopher Fine.
New York: Atheneum, 1989.

Senses, instincts, and ways of behaving, along with characteristics such as color and shape, help all creatures survive in the wild. This book offers fabulous candid pictures of sea creatures at home in the ocean. Discover interesting tidbits of information as to how these wonderful animals fit into their underwater environment.

***What's for Lunch: The Eating Habits of Seashore Creatures* by Sam and Beryl Epstein.**
New York: Macmillan, 1985.

What do sea animals eat? This book describes not only what the animals eat for their meals, but how they find their food. You'll also discover the tricks some creatures use to hide from enemies searching for dinner.

***Where the Waves Break: Life at the Edge of the Sea* by Anita Malnig.**
Minneapolis: Carolrhoda Books, 1985.

Explore the living things that make their home along the coast. Some animals hide in the sand and rocks and others in small pools of water. Read this book to find out where you can find some of these plants and animals.

Theme T STABILITY

"Water, water, every where,

Nor any drop to drink."

—from "Rime of the Ancient Mariner"
by Samuel Taylor Coleridge

Salty Seas

This is a description of what it's like to be in a boat in the middle of the ocean. Most of Earth is covered by oceans. With all the oceans on Earth, why would you not want to drink water from them?

Did you know that it is easier to swim in ocean water than it is to swim in a lake or swimming pool? Water from the ocean is different from the water found in rivers, lakes, ice, or a pool. Ocean water is salty and not only tastes bad but can also make a person ill. Have you ever tried to float in salt water? Do the following activity to see what happens to an egg in fresh water and in salt water.

TRY THIS Activity! Does It Sink or Does It Float?

What You Need

egg, salt, 12-oz. jar, water, small spoon, *Activity Log* page 3

Fill the jar half full with water and put the egg in the jar. Record what happens to the egg. Remove the egg and stir 1 spoonful of salt into the water. Replace the egg and observe.Continue to add 1 spoonful of salt to the water until you notice a change in the egg. Write your observations in your ***Activity Log.***

Salt and other materials give ocean water different properties from other water and make it easier to float in the ocean. In this lesson you will explore the composition of ocean water.

Atlantic Ocean

EXPLORE

Activity!

How Can You Layer Liquids?

In this activity you will investigate the effect that different amounts of salt have on water. You will also compare salt water with fresh water. As you do, think about how ocean water is different from fresh water.

What You Need

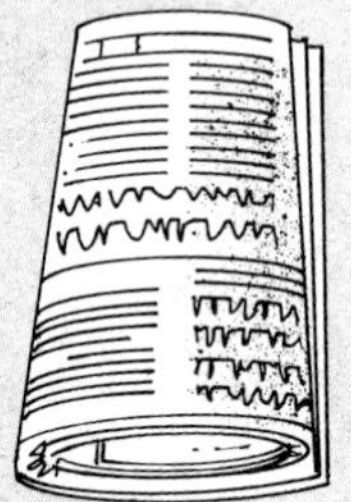

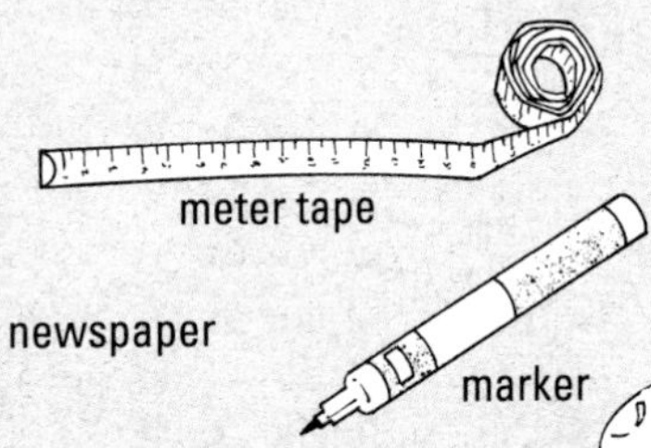

What To Do

1. Spread the newspapers on your desk. Fill four of the cups half full of water.

2. Add yellow food coloring and 1 spoonful of salt to one of the cups. Add red food coloring and 2 spoonfuls of salt to another cup. Add blue food coloring and 3 spoonfuls of salt to the third cup.

3. From the bottom of the straw, make a mark every centimeter for 4 cm.

4 Gently place the bottom of the straw 1 cm below the surface of the blue liquid. Seal the top with your finger. Lift the straw out of the cup.

5 With your finger still on top of the straw, place the bottom of the straw 2 cm down in the clear liquid. Lift your finger off the straw and then put your finger back on the straw. Lift the straw out of the water.

6 Observe the water in your straw. Draw and label your observations in your ***Activity Log.*** Put the liquid in the straw into the empty cup by releasing your finger from the top of the straw.

7 Continue experimenting with the other liquids in any order.

8 Use the data collected in your observations to make layers with all four.

What Happened?

1. What two things happened when you put two different liquids in the straw?
2. How did you arrange the colors of the liquids from the bottom of the straw to the top of the straw?

What Now?

1. Why do some of the liquid combinations mix in the straw while others form layers?
2. Which is heavier, equal amounts of fresh or salt water? Why?
3. How did the different amounts of salt change the properties of the salt water?
4. In how many different orders can you arrange the liquids so that layers form?

What's in Ocean Water?

In the Explore Activity, you explored the property of salinity. **Salinity** (sə lin´ i tē) is the amount of dissolved salt in water. You discovered that water with a lower salinity (less salt) will float on water with a higher salinity (more salt). If you put the water with the highest salinity on the bottom of the stack, it stays on the bottom. If you try to put the water with the highest salinity on the top of the stack, it sinks through the less salty water and mixes with it, so no layers form.

Salt

Ocean water contains more minerals and gases than fresh water. Have you ever tasted ocean water? Since the most abundant mineral dissolved in ocean water is salt, it tastes salty. There is so much salt in the oceans that if all of it could be taken out, it would form another continent of solid salt the size of Africa!

At least six million tons of salt are obtained from ocean water every year. Most of this is used for snow removal, water softening, and refrigeration.

Health Link — *Water That Makes You Thirsty*

The human body needs some salt to function properly. However, too much salt can be dangerous to your health. Have you ever felt thirsty after eating salty food such as chips or pizza? The reason is your body uses water to rid itself of extra salt. When you eat a lot of salt, your body needs a lot of water, which is why you feel thirsty. If you drank only ocean water, your cells would eventually dry out as they used up water trying to get rid of the extra salt. Drinking more ocean water would add more salt but no fresh water for your body to use. The continued loss of water from your body would make you very ill.

Other Minerals

Besides salt, ocean water contains all of the elements that occur naturally, and every mineral found on Earth. Some minerals are separated from ocean water and used in industry. There is enough of the element gold in the oceans for every person on Earth to have nine pounds. However, the gold particles are so tiny that it is very difficult and expensive to separate them from the other elements and minerals.

Gases

Gases are also dissolved in ocean water. The same gases in the air you breathe are in the ocean. Gases dissolved in ocean water include oxygen, nitrogen, and carbon dioxide. Plants and animals that live in the ocean need these gases to survive.

Gas bubbles are easily seen in this water.

Salt in the Oceans

How did all the salt get in the oceans? The salt in the oceans comes from minerals in the rocks on Earth.

1 *Rain and water break down the minerals.*

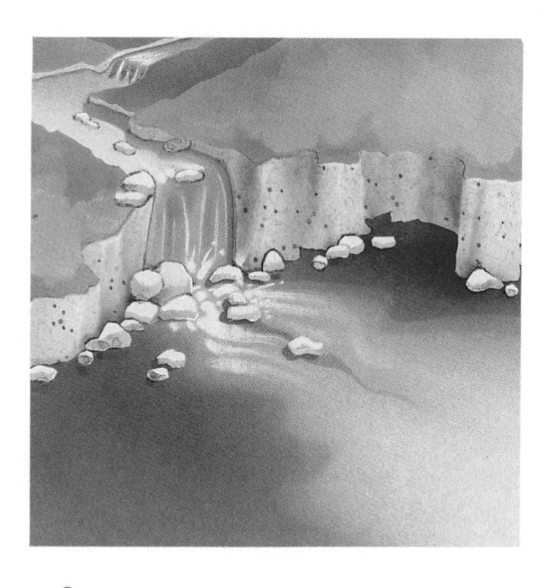

2 *The rivers carry the dissolved minerals and salt to the oceans. They mix with the water in the oceans and the oceans become salty.*

Water carries salt from the rocks to the oceans. When the heat from the sun evaporates water from the oceans, more and more salt gets left behind. As time passes, the salinity of the oceans should increase. But does it?

The salinity of the oceans does not seem to be increasing. Scientists hypothesize that some salts are removed from the water as fast as they're added. Do the Try This Activity on the next page to see how much salinity does change.

Plants and animals use salt to build shells and skeletons.

TRY THIS Activity! Salinity—Higher or Lower?

What You Need

12-oz. jar, water, salt, small spoon, *Activity Log* page 6

Make a saltwater solution in the jar. Allow the water to evaporate over a period of a few days. Should the salinity of the remaining water be higher or lower? Why? If you added more salt water, how would the salinity change as the water evaporated? Explain the results in your ***Activity Log.*** Why is this different from what happens in ocean water?

The movements of the ocean generally keep ocean water mixed so that the salinity is almost the same throughout. But there are places where you can see some differences in salinity, as you did in the Explore Activity.

At the mouth of the giant Amazon River in Brazil, the fresh water from the river fans out into the ocean, forming a layer over the ocean's salt water.

Social Studies Link *Lakes Have Salt, Too!*

Not all of the salt water on Earth is in the oceans. Choose one of the following salt lakes and salt seas and locate it on a map or globe: Great Salt Lake, Utah; Pyramid Lake, Nevada; Salton Sea, California; Albert Lake, Oregon; Devils Lake, North Dakota; Lake Pontchartrain, Louisiana. Research the composition and uses of the lake or sea and share your information with the class.

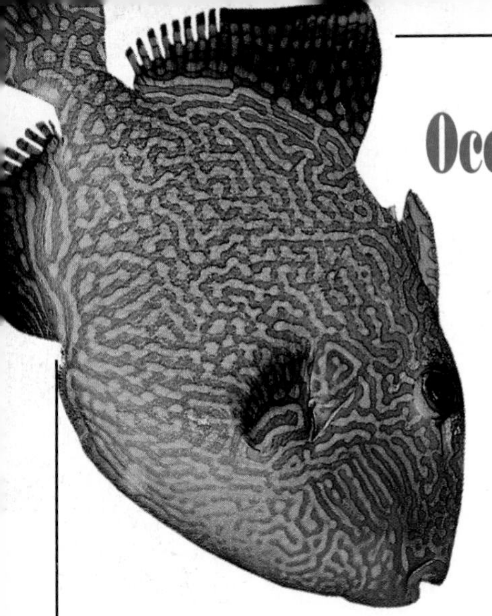

The triggerfish shoots water at the sea urchin to blow it over. The triggerfish can use its sharp teeth to eat the unprotected sea urchin.

Ocean Animals

Ocean animals have characteristics that allow them to survive. Ocean animals have also adapted to the high level of salt in their environment. For example, saltwater fish drink a lot of water because they lose body fluids through their skin and gills. They need and use all the water they drink. Freshwater fish are different. They absorb, or take in, water through their skin and gills. Therefore, freshwater fish don't need to drink any water. Some fish, like salmon, smelts, and butterfly mudskippers, can adapt to live in fresh water or in salt water. However, very few fish can make this adjustment.

Porcupine fish

The purple sea urchin of the North American Pacific coast is shaped like a ball and is covered with long movable spines. The sharp spines keep other animals from eating the sea urchin.

Minds On!

Imagine you are an organism that lives in the ocean. How do you look? What do your surroundings look like? How do you move? How do you get your food? When another living thing tries to eat you, how do you escape? Use page 7 of your ***Activity Log*** to record your answers.●

Sum It Up

Without oceans, there would probably be no life on Earth. Though we don't drink salt water, we use the salt and the other minerals we get from the oceans. Salt and other dissolved minerals flow into the oceans from rivers. Though salt is constantly added to the oceans, plants and animals use salt, so the salinity stays the same.

In the next lessons, you'll explore how important the oceans are to you, too.

Critical Thinking

1. The planet Mars has ice but no oceans. How would that fact alone make life on Mars different from life on Earth?
2. Rivers carry minerals and salts to the oceans, so why aren't rivers salty?
3. Why would you expect the salinity of the oceans to rise with time?

The lionfish has long fancy fins that contain a deadly poison. The lionfish uses these deadly fins to attack other fish and protect itself.

The soft-bodied octopus has eight arms, called tentacles, that it uses to capture food.

The porcupine fish fills itself with water so that it looks bigger to other fish.

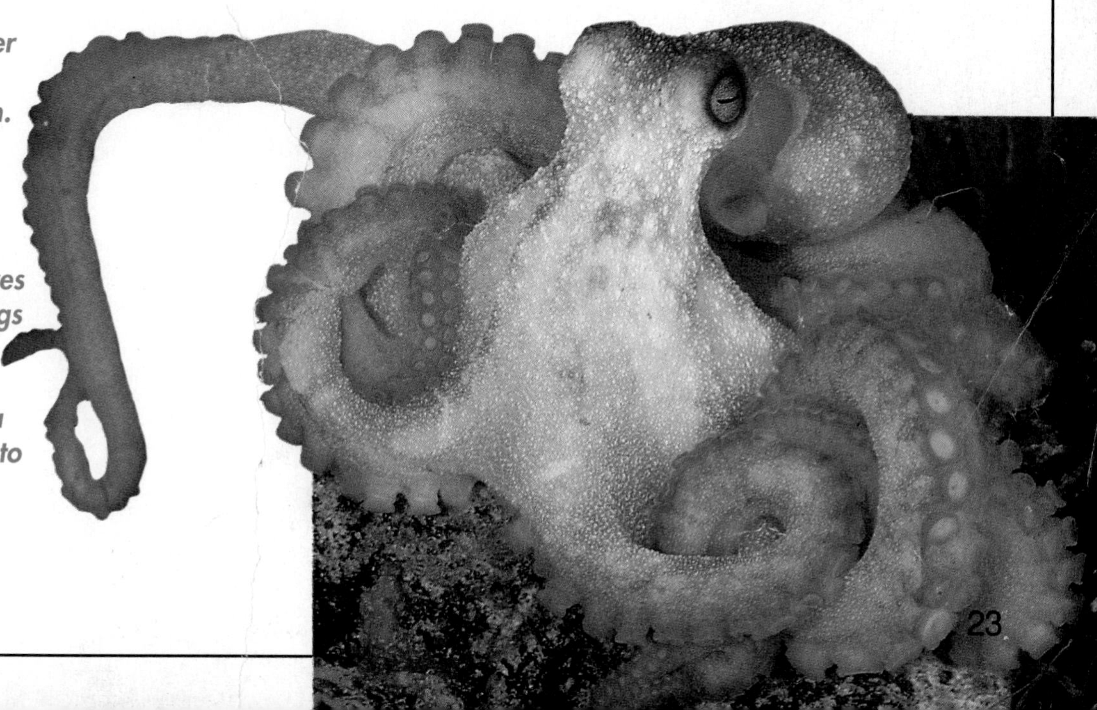

The fiddler crab lives on beaches and digs into the sand. The male has a huge front claw, called a pincer, that it uses to fight with other males.

Movement of Ocean Water

The waters of the ocean never stop moving. Even in calm seas, ocean waters rise and fall. Water moves out of the oceans, too. It moves into the air, onto the land, into the rivers, and back into the oceans.

In the last lesson, you discovered that ocean water is different from fresh water. Ocean water contains dissolved minerals and salts, which make it salty. Some ocean water has more salt than other ocean water. There are other differences as well. Ocean waters near the equator are warm, while polar waters and deep ocean waters are colder. Differences in temperature are one reason ocean water moves. In this lesson, you will explore the ways that ocean water moves over Earth.

Indian Ocean

EXPLORE

Activity!

Do Hot Water and Cold Water Mix?

The temperatures of ocean waters are not always the same. Some areas of the ocean have warmer waters than others. In this activity you will see how differences in temperature affect water. You will also see what happens when ocean areas with two different temperatures meet.

What You Need

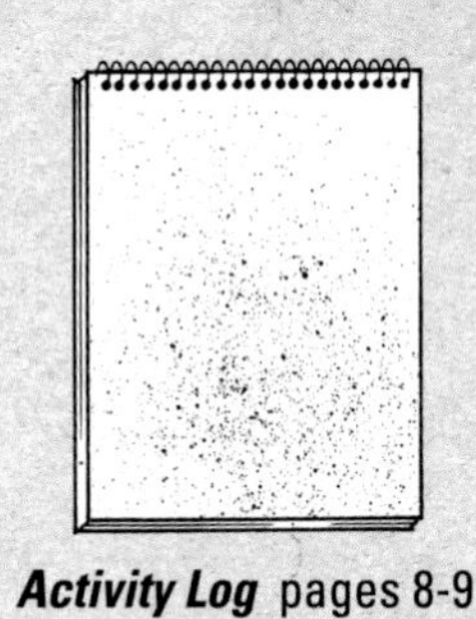

Activity Log pages 8-9

blue ice water

red hot water

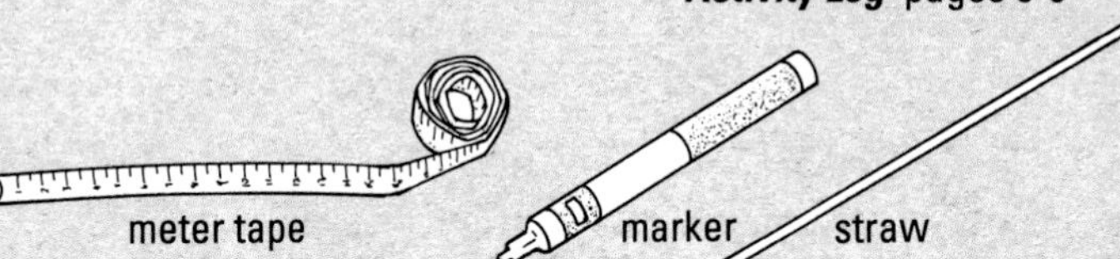

plastic cups

What To Do

1. Make marks 5 cm and 10 cm from the bottom of the straw.
2. Fill one cup almost to the top with blue ice water. Fill another with red hot water.
3. Place the straw down to the first mark in the blue water. Put your finger over the top of the straw. Lift the straw out of the water.

4 With your finger still on the straw, put the straw down to the second mark in the red water. Lift your finger off the straw so that the water in the straw will rise. Place your finger back over the top of the straw and lift the straw out of the water.

5 Observe the two colors of water in the straw. Record your observations in your ***Activity Log.***

6 Place the straw in the empty cup. Lift your finger to release the liquid.

7 Have another member of your group repeat steps 3–5, placing the straw in the red water first and then in the blue water.

8 With your finger still on the straw, slowly tilt the straw so that it is almost horizontal. Observe what happens.

What Happened?

1. What happened when you put the hot water in the straw first?
2. What happened when you put the cold water in the straw first?
3. What happened when you tilted the straw?

What Now?

1. Which water is heavier? Give reasons for your answer.
2. Why did the water react differently depending on the order of the cold water and the hot water in the straw?
3. What would you expect to happen if deep ocean water became warmer than the water above it? Why?

Looking Back

How is this activity similar to the activity you did with salt water in Lesson 1? Using what you discovered in these two activities, list the two factors that determine how heavy ocean water is.

Water Movement

In the Explore Activity, you discovered that hot water floats on top of colder water. As you noticed, hot water and cold water do not mix easily. When hot and cold water meet, layers form just as they did in the salt water activity. When you put the cold water on the top, it sank through the hot water.

Surface Currents

One way ocean water moves is by currents (kûr´ ənts). A **current** is a strong flow of water. Winds add to the water movement in the ocean by causing **surface currents.** If you blow across the surface of a bowl of water, you can make the water move. This is similar to how winds cause currents on the ocean's surface. Surface currents carry cold water along one coast or warm water along another coast. The temperature of the water affects the weather over the ocean and land.

At the equator, most currents flow westward until they bump into a continent. Then the currents turn either north or south.

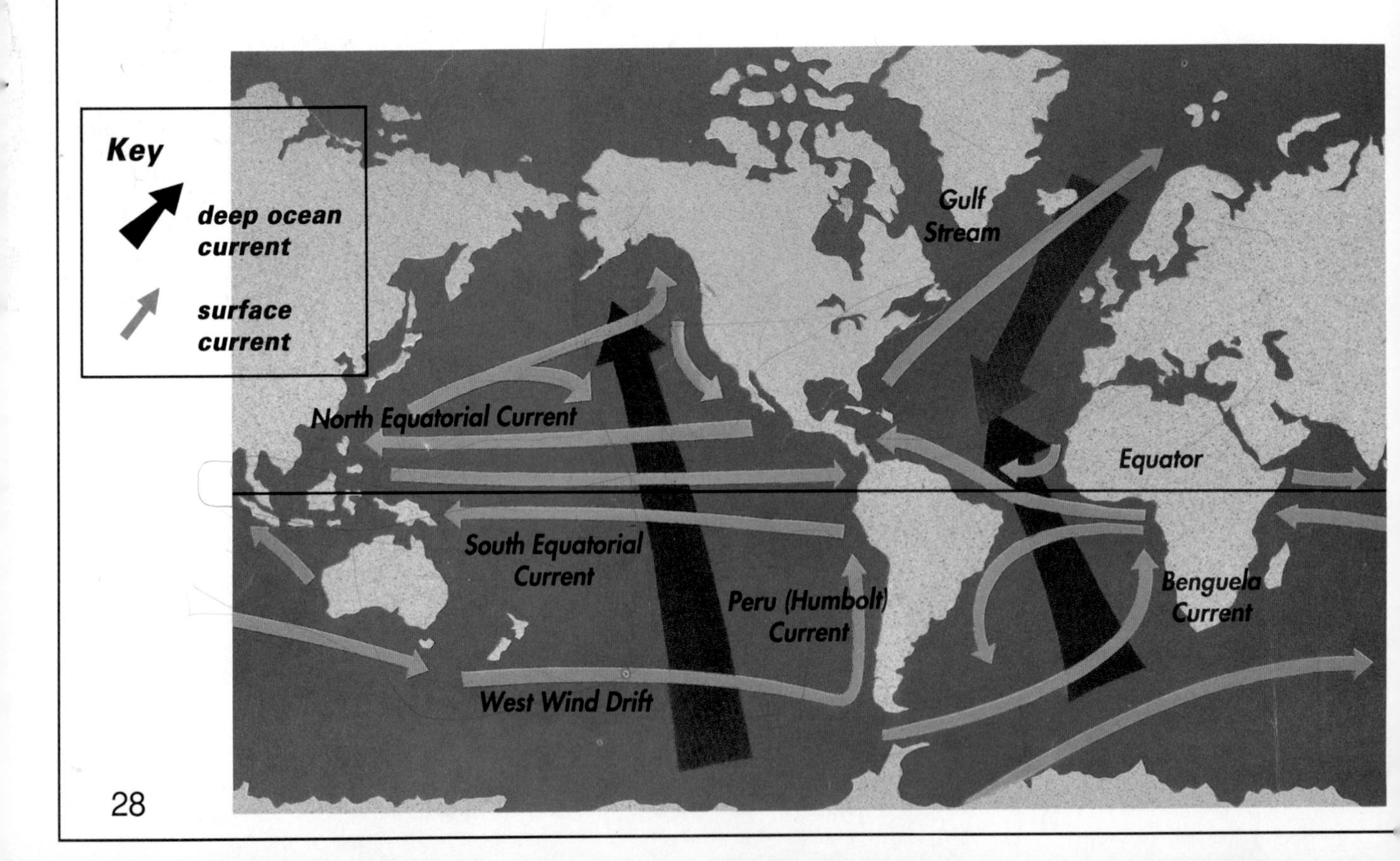

Deep Ocean Currents

Differences in temperature can also cause ocean water to move. Colder water sinks in the ocean just as it did in the straw. These movements cause **deep ocean currents.** Deep ocean currents can occur many kilometers below the surface of the ocean and they move very slowly. They usually flow in the direction opposite to surface currents.

Waves

An **ocean wave** is the up-and-down movement of ocean water caused by the energy of the wind. When wind blows across the surface of the ocean, ripples form and grow into waves. The size of a wave depends on the wind's speed, distance, and length of time over the ocean's surface. Ocean waves are very powerful. Waves can pick up large rocks and throw them up onto the shore.

The constant pounding of waves wore a tunnel-like hole in this rock.

Literature Link *Creatures of the Sea*

How might the movements of the ocean affect the creatures living in it? As you read *Creatures of the Sea,* think about how each animal moves in the water. Would strong waves and currents be helpful or harmful to its existence?

TRY THIS Activity! Making Waves

What You Need

rope, masking tape, *Activity Log* page 10

You can use a rope to understand how waves move through the ocean. Place a piece of tape on the middle of the rope. Start with the rope held loosely between you and your partner. One person shakes the rope while the other holds the opposite end. Describe the movement of the tape on the rope. How does the movement of the rope compare to waves in an ocean?

The water in an ocean wave moves like the tape on the rope. As a wave goes across the ocean, the water moves up and down.

Tides

Ocean water is always moving. Surface and deep ocean currents push it along. Waves rush through it. The ocean also moves in a third way, with the tides. The rising and falling of ocean water levels is called a **tide.**

High tide at Bay of Fundy, Nova Scotia

Low tide at Bay of Fundy, Nova Scotia

Tides are caused by gravity. **Gravity** is the pulling force between two objects. Ocean tides are caused by the pull of gravity between the moon and Earth. Tides are also influenced by the pull of gravity between the sun and Earth.

Tides follow the movement of the moon around Earth. The moon pulls the nearest ocean water toward it, causing the ocean to bulge. At the same time, another high tide bulge forms on the opposite side of Earth. This second bulge occurs because the moon pulls on Earth more than it pulls on the ocean.

Harbor activity depends on the tides. Some ships are too big to pass through a shallow harbor entrance at low tide. At high tide, when the water is deeper, the ship travels through easily.

Coastal cities have tide tables that show the times for all high and low tides during the year. Ships' captains use these tide tables to determine when it is safe to pass through a harbor and when it isn't.

Harbor Pilot

Large ships go from port to port. At each port, a ship goes from deep ocean water through the more shallow water of the harbor to the dock. The ship's captain needs the services of a harbor pilot to guide the ship through the harbor. In a harbor, a large ship must seek the deepest water and the safest route. If a ship drifts from the major ship route, it might hit rocks or an old sunken ship. It could get stuck on the bottom.

Harbor pilots study the depths of the water, the effect of the tides on the harbor, and the dangers in a port area. These trained pilots know how to follow floating markers called buoys to guide a ship safely to the dock.

Most harbor pilots have 30 to 40 years of experience. Some have served an apprenticeship. All pilots must have good vision. A pilot must pass a test to get a license as a harbor pilot.

Oceans and the Water Cycle

Another important movement of ocean water is the water cycle. Because the oceans cover so much of Earth's surface, most evaporation occurs over the oceans. When ocean water evaporates, the water vapor moves into the atmosphere and the salt is left behind. So even though the rain you see and feel during a thunderstorm came from the salty ocean, it doesn't taste like salt water. The water cycle provides the fresh water that plants and animals on Earth need to survive.

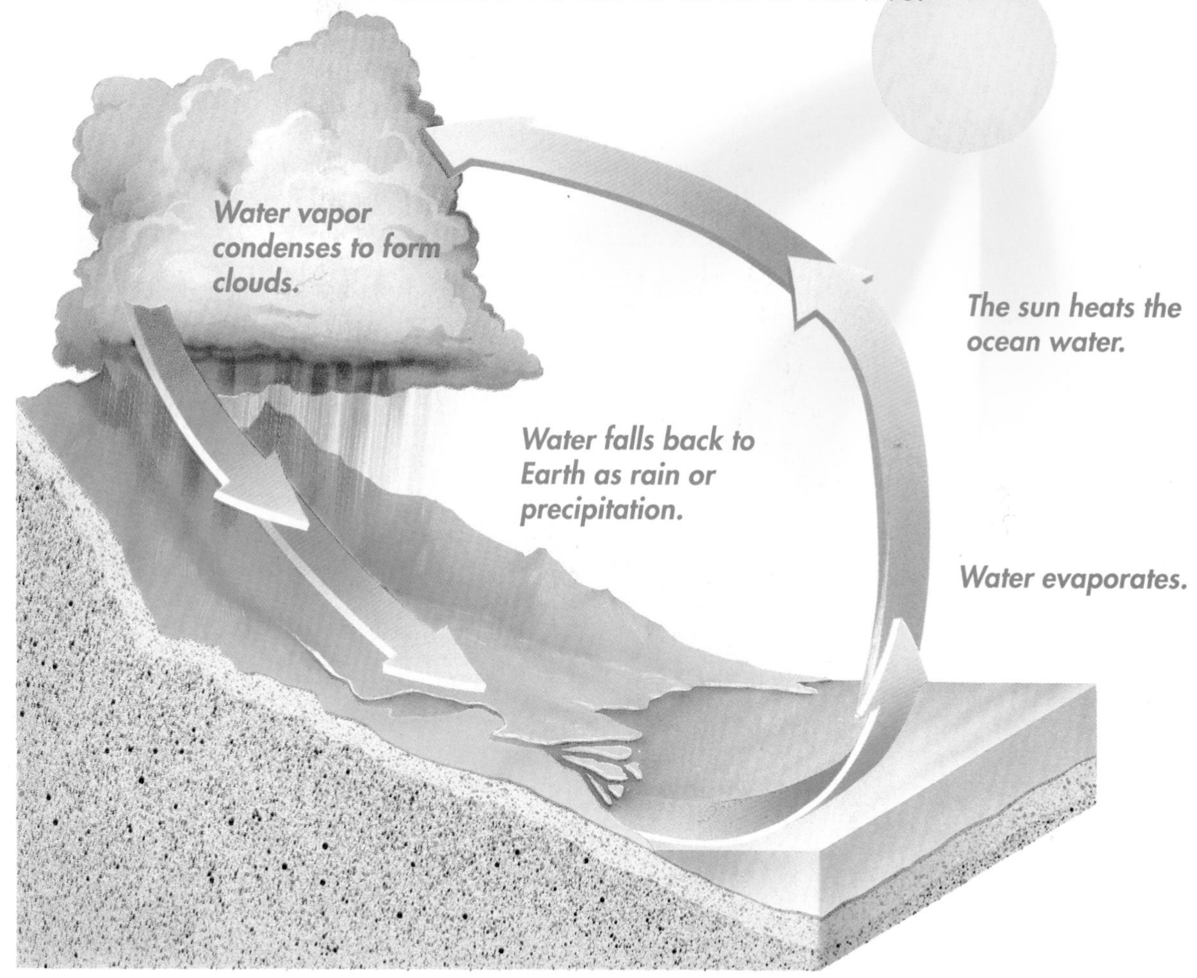

Minds On! Pretend you are a drop of water in the ocean. On page 11 in your ***Activity Log,*** write or draw a story about your trip through the water cycle.●

The Movement of Ocean Life

Some types of ocean life depend on the movement of ocean waters. Animals that do not have the ability to swim float with the currents. They rely on the movement of ocean waters to move them from place to place. Some animals attached to the floor of the ocean are filter feeders. They depend on the movement of ocean waters to bring them food.

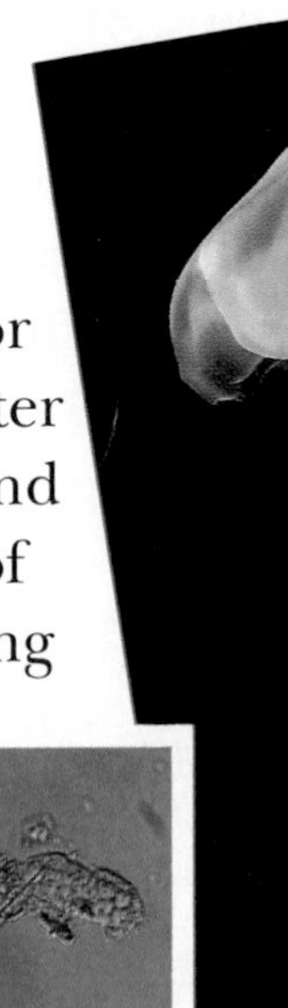

Jellyfish float with the currents.

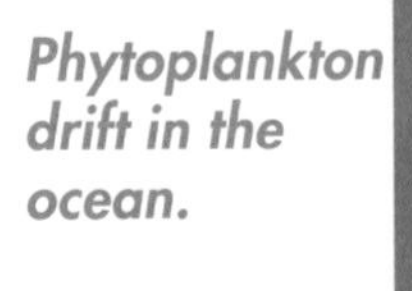

Phytoplankton drift in the ocean.

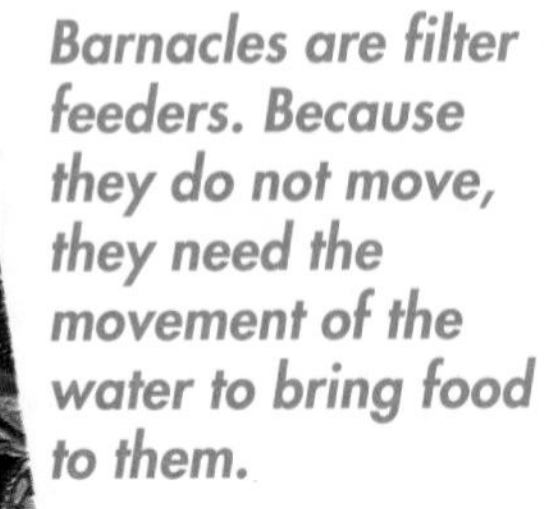

Barnacles are filter feeders. Because they do not move, they need the movement of the water to bring food to them.

Sum It Up

The ocean is a large moving system. Interactions between ocean waters of different temperatures may produce currents. Winds may cause currents on the surface of the ocean. Winds also blow water into waves. Waves carry energy through the ocean system. The interaction between the moon and Earth causes the tides to rise and fall each day.

Ocean water also interacts with the atmosphere. As the sun heats up the ocean, water evaporates and enters Earth's atmosphere. Clouds form all over Earth and rain carries water back to the oceans, rivers, and land.

Angelfish

Fish swim to find food, to escape or hide from enemies, or to move to a better location.

Critical Thinking

1. If you built a sand castle on the beach at low tide, what might happen to it hours later? Why?
2. Imagine you're floating on a raft in the middle of the ocean. How will the currents affect you?
3. Explain how each type of ocean water movement affects ocean life along the shore.

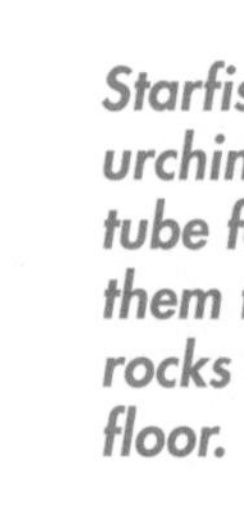

Starfish and sea urchins have tiny tube feet that allow them to move over rocks and the ocean floor.

The Shape of the Ocean Floor

Sebastian the crab from Disney's film **The Little Mermaid** *says, "It's better down where it's wetter." Do you think so, too? How deep is the sea? What does it really look like?*

Even if you haven't visited the ocean, you've probably seen photos of it. Usually the ocean water looks blue. Perhaps a graceful boat sails across the water close to the shore. People wade and surf in the ocean waves. They fish from the shore or from boats. Using goggles and air tanks, some people scuba dive to view the sea life beneath the surface.

Near the shore, it can be easier to see the bottom of the ocean. But away from the shore, the ocean gets deeper. The deep parts of the ocean floor are dark, where natural light can't reach, and hard to get to.

Minds On! With your group, use a globe to look at the oceans. Choose an area of an ocean and think about the ocean bottom and how it might look. Then, on chart paper, take turns drawing the side view of a feature this part of the ocean might have. Draw a smaller version on page 12 in your ***Activity Log.*** ●

Pacific Ocean

EXPLORE Activity!

Mapping the Ocean Floor

It isn't easy to map the ocean floor. People cannot easily see its shape as they can with land. As you complete this activity, think about ways to map something you can't see.

What You Need

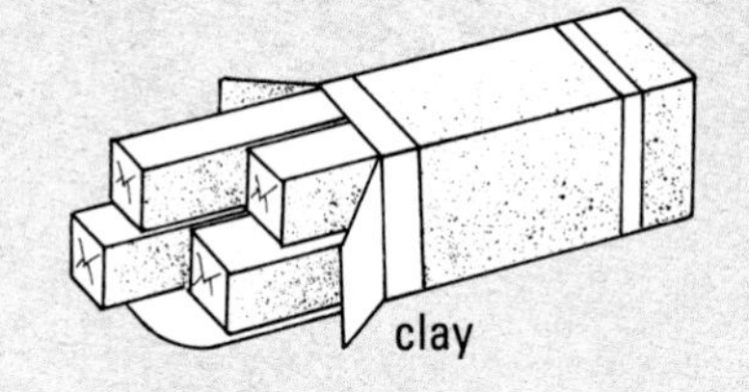

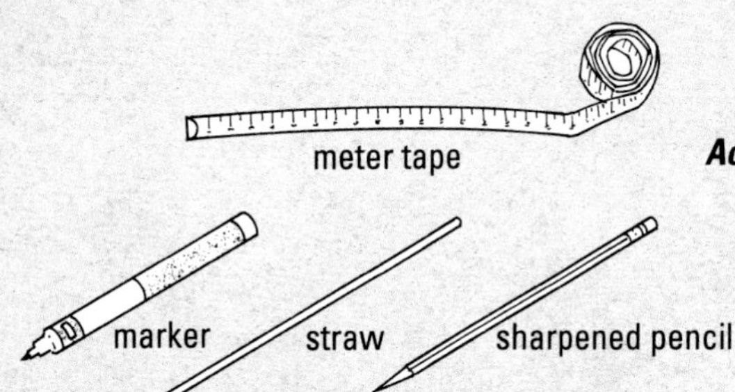

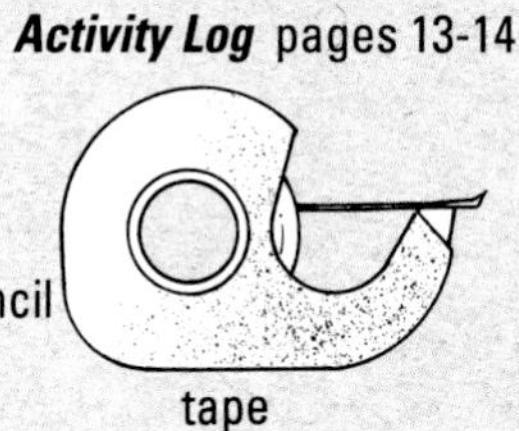

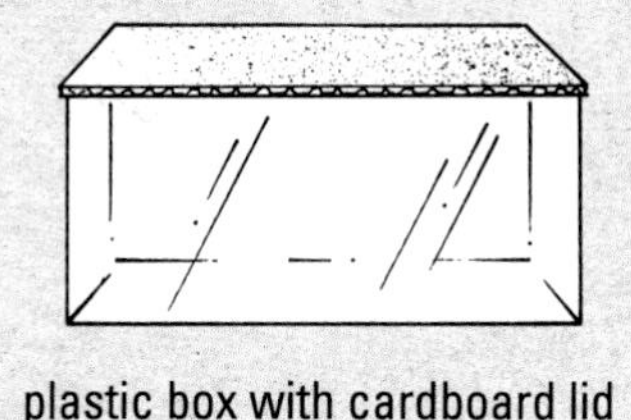

What To Do

1 Make a mark every 3 cm on the top of the box in a straight line. Using a very sharp pencil, poke a hole at every mark large enough for the straw to fit through. Number the holes. ***Safety Tip:*** As you make the holes, point the pencil away from yourself and those nearby.

See the *Safety Tip* in step 1.

2 Use the clay to make an irregular, or lumpy, surface on the bottom of the box.

3 Tape the lid onto the box and trade boxes with another group.

4 Use the ruler to make marks on your straw every 1 cm.

5 Plan a way to use the straw as a depth gauge. Use your depth gauge in each hole and record your findings on the data table in your ***Activity Log.***

6 Using the marks on your data table, draw a profile of the surface of the clay.

What Happened?

1. What was your straw measuring when you poked it through the holes in the shoe box?
2. Take the lid off the box and, looking through the side, make a drawing of the shape of the clay bottom. Compare it to your line drawing.

What Now?

1. Why would it be important to know the depth of the ocean along the shore?
2. How could you use a technique similar to the one you devised in this activity to map the ocean floor?

Under the Sea

In the Explore Activity, you measured the depth of the box with a straw and recorded each measurement on a graph. When all the points on the graph were connected, it gave you a picture of how the clay floor in the box looked. In earlier times, scientists used weighted ropes in a similar way to measure the depths of the ocean.

Oceanographers (ō´ shə nog´ rə furz) are scientists who study the ocean. They have produced maps of the ocean floor using sonar. **Sonar** (sō´ när) is an instrument that uses sound waves to find underwater objects. Do the Try This Activity on this page to see how this works.

TRY THIS Activity!

Sonar Mapping

What You Need

high-bouncing ball, clock or watch with second hand, *Activity Log* page 15

You can do sonar sounding using a high-bouncing ball. Drop the ball onto a flat, hard surface while your partner times from when the ball leaves your hand until it returns to your hand. From a different height, drop the ball onto the flat, hard surface again and catch it, recording the time. Take turns with your partner, dropping the ball from various heights and recording the times. What effect does distance have on the time of the ball's return? Record your results in your ***Activity Log.***

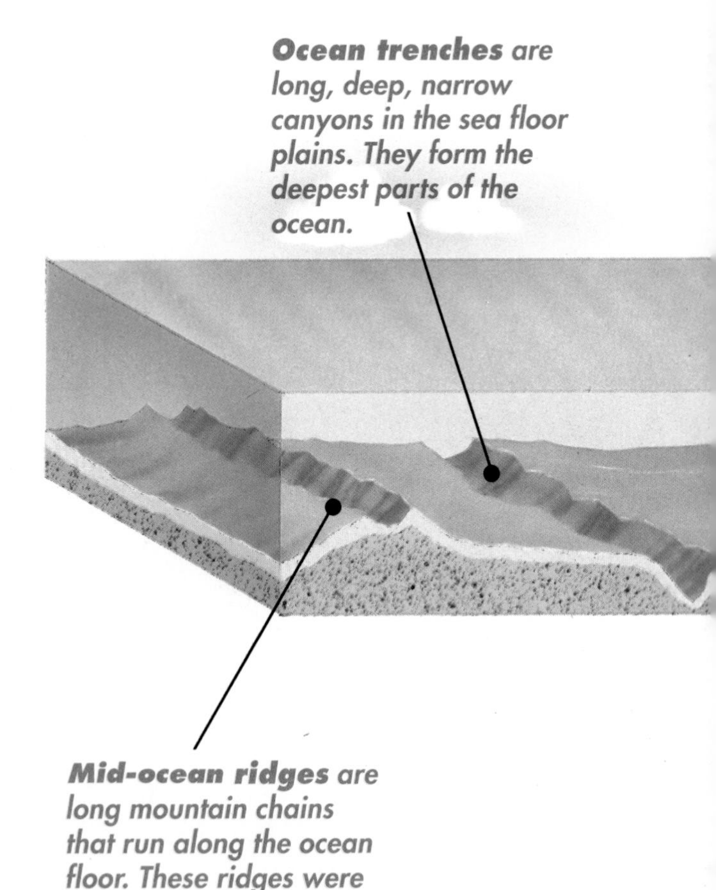

Ocean trenches *are long, deep, narrow canyons in the sea floor plains. They form the deepest parts of the ocean.*

Mid-ocean ridges *are long mountain chains that run along the ocean floor. These ridges were formed from volcanic activity under the ocean floor.*

The high-bouncing ball is like a sound wave. The sound travels through the water to the sea floor and bounces back to the ship. The longer the sound wave takes to return to the ship, the deeper the water. This method has been used to map many areas of the ocean.

The oceans have low, flat plains that are somewhat like Earth's land areas. However, the oceans have longer mountains and deeper trenches than any found on Earth's surface.

*Large volcanoes rising out of the sea floor plains are called **seamounts.** Some seamounts rise high enough to break the surface of the ocean water and become islands.*

*The **continental slope** starts at the edge of the continental shelf and slopes down to the deep ocean floor at a steeper angle than the continental shelf. Deep canyons cut through the slope.*

*The **continental shelf** is the gently sloping edge of the continent that is covered by ocean water.*

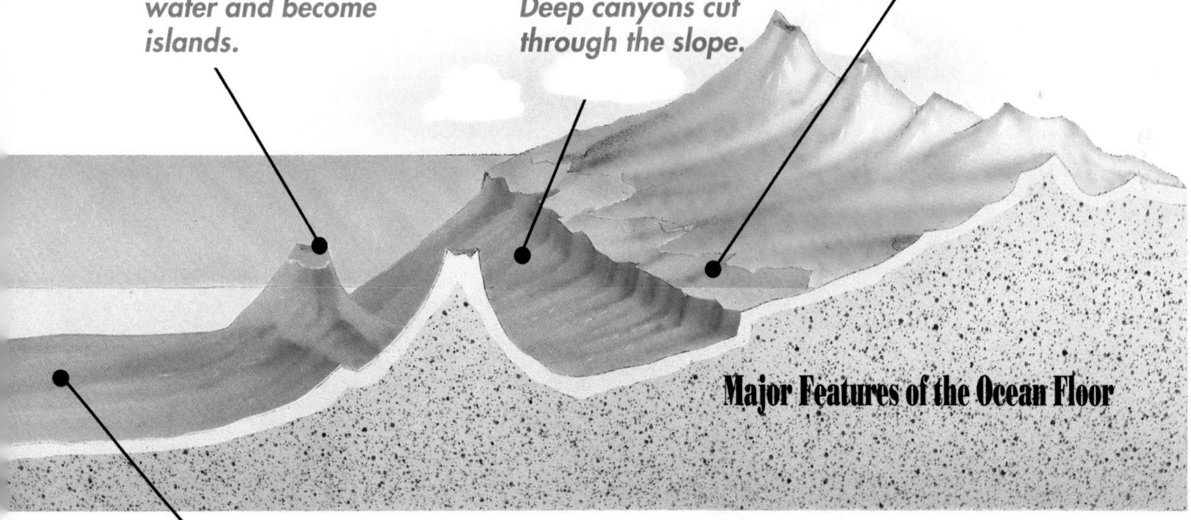

Major Features of the Ocean Floor

*The **seafloor plains** are large regions of the sea floor that appear flat.*

Music/Art Link *Add Features to Your Profile*

Refer back to the Minds On in which you drew a side view of one area of an ocean. Since you made that drawing, you've learned more about the ocean floor. Add new features to your drawing of the ocean floor.

Exploring the Ocean Depths

Early attempts to explore beneath the sea were limited by a diver's need to breathe. The equipment used received air and power from lines attached to a ship. If those lines were cut or damaged, power and air was lost.

Even in a bathtub full of water, you can feel the water pressing on you. The deeper the water, the more the pressure of the water increases. When people want to explore deep in the ocean, they not only have to breathe, they also have to protect themselves so that the pressure does not crush them.

New, one-atmosphere diving suits provide the diver with air and protection from extreme water pressure.

Old diving suits used air hoses attached to the boat.

Literature Link *Creatures of the Sea*

Choose an animal from *Creatures of the Sea* and investigate how it uses camouflage. Would your animal be able to use its "disguise" to live successfully in another area of the ocean?

Focus on Technology *Submersibles*

Today, most underwater research is carried out by submersibles. **Submersibles** are underwater machines that can travel freely around the ocean waters. They not only move up and down, they can also maneuver in all directions. Though some submersibles carry people, many do not. They are operated by remote control. This means they can be operated from a distance, such as from on board a ship.

Submersibles helped find the wreck of the ship *Titanic*. The *Titanic*, a famous ocean liner, sank in 1912 when it hit an iceberg. In 1985, two remote-controlled submersibles located the *Titanic* on the floor of the Atlantic Ocean. Television cameras aboard the submersibles sent back pictures of the wreckage.

*A **bathyscaphe** is an underwater boat that has its own power source and air supply. In 1960, a bathyscaphe called the Trieste descended 10,910 meters (35,800 feet) into the Mariana Trench, the deepest part of any ocean.*

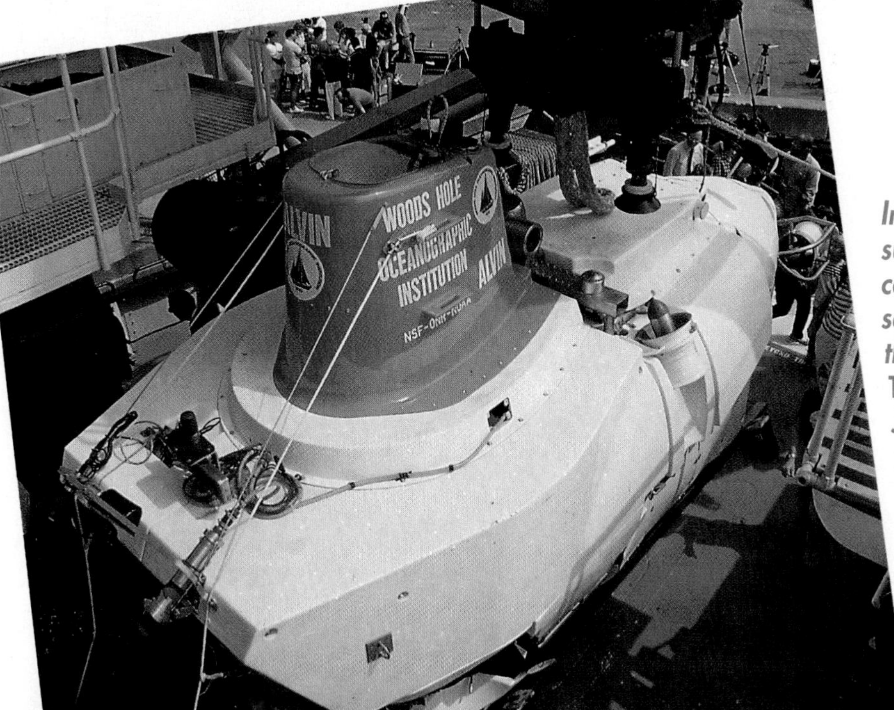

In 1986, the submersible Alvin carried people to the sea bottom to explore the wreckage of the Titanic. A robot called Jason Jr. roamed through the wreckage taking pictures.

Life in the Deep Ocean

Until recently, we knew little about living things in the depths of the oceans. Organisms of the deep ocean live under conditions different from the living conditions of other organisms. People once thought that nothing could survive in the waters of the deep ocean. No one believed that organisms could live with the extreme water pressure and the lack of light in the deep ocean waters.

Bioluminescence *is the ability of an organism to produce its own light. Most bioluminescent animals live in the deep ocean.*

The deep-sea hatchetfish has rows of light organs along the lower edge of each side of its body.

The tip lights of the deep sea anglerfish attract its next meal.

Sum It Up

Would you like to become a scientist and explore the ocean floor? That's not an easy job because much of the ocean is very deep and dark. Yet, using sonar, scientists have created maps of the ocean floor. Using those maps, perhaps you could drive a submersible to the bottom of the ocean. You would see undersea mountains, ridges, and volcanoes along with flat plains and the edges of the continents. You could compare the ocean floor with areas of Earth's surface that are more familiar to you.

The gulper eel can open its large jaws to catch prey larger than itself.

Critical Thinking

1. Why would people want to explore the bottom of the ocean?

2. How can sonar tell the difference between a mid-ocean ridge and a trench?

3. How do you think underwater mountains are different from mountains on land? How are they similar?

Pineapple fish stay on the ocean bottom during the day and travel toward the surface at night.

People and the Ocean

If there were no oceans on Earth, there would probably be no life on Earth. Even if you live far away from the ocean, you depend on it more than you think.

Everything that lives on Earth consists in part of water. Your body is about 66 percent water. An elephant is 70 percent water, an earthworm is 80 percent water, and a juicy red tomato is about 95 percent water. These living things all need water to live and grow. In this lesson you will discover ways we depend on oceans you might not be aware of.

Minds On! Water is important to all of us. With a partner, brainstorm a list of the ways that you've used water today. Record your ideas in your ***Activity Log*** on page 16. ●

Arctic Ocean

EXPLORE **Activity!**

How Can You Make Fresh Water From Salt Water?

Of all the water on Earth, 97 percent of it is in the ocean. Only 1 percent of the remaining water is available for use as drinking water and water for crops. Fresh water is a scarce resource. It will become even more important in the future.

What You Need

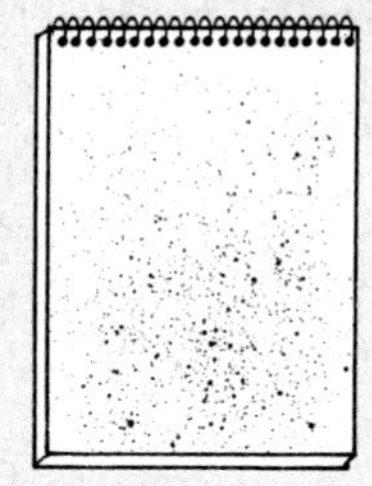

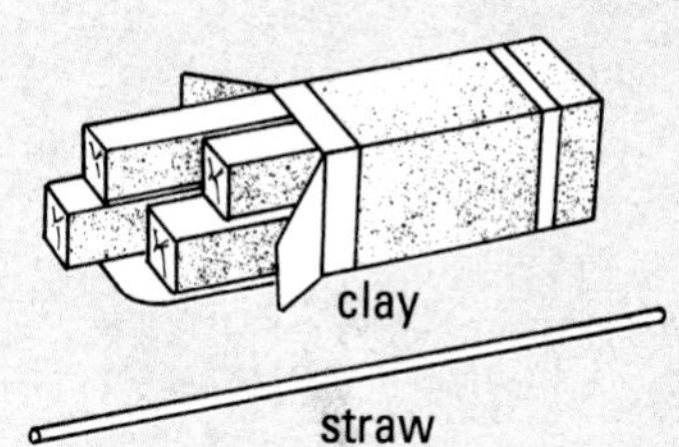

safety goggles

rubber band

plastic wrap

What To Do

1. Place a piece of clay in the bottom of the bowl.

2. Stir 2 spoonfuls of salt into the container of water. Pour salt water into the bowl to a depth of about 2 cm.

3. Using the straw as you did in the Explore Activity on pages 16–17, taste the salt water. Describe what you tasted in your ***Activity Log.*** Rinse the straw with fresh water. *Safety Tip:* Never taste anything in an experiment unless you are told to do so.

See the *Safety Tip* in step 3.

4 Put the small container on the clay in the center of the bowl of salt water.

5 With the safety goggles on, put plastic wrap over the bowl. The wrap should hang down over but not touch the salt water. Hold the plastic wrap in place with the rubber band.

6 Put a clay ball the size of a marble on top of the plastic wrap so that it hangs directly over the small container.

7 Place the container in a sunny location for one day.

8 The next day, write your observations in your ***Activity Log.***

9 Use the straw to taste the water in the small and large containers. Record your observations in your ***Activity Log.***

What Happened?

1. What happened under the plastic wrap?
2. Compare the tastes of the different water samples.

What Now?

1. Why did one sample of the water taste salty while the other did not?
2. How can ocean water be changed to water that we can drink?
3. How could you speed up the process and produce more fresh water from salt water?

How We Use the Ocean

In the Explore Activity, you found a way to change salt water to fresh water. Water evaporated from the saltwater mixture, leaving salt water behind. Then it condensed on the plastic wrap and fell into the small container as precipitation. So when you tasted the water in the small container, you tasted fresh water and not salt water. You created a water cycle inside a bowl!

As the population of the world increases, so does the need for drinking water. In some parts of the world, such as Saudi Arabia, there is not enough fresh water. People can't survive without fresh, drinkable water.

Desalination plant in Freeport, Texas

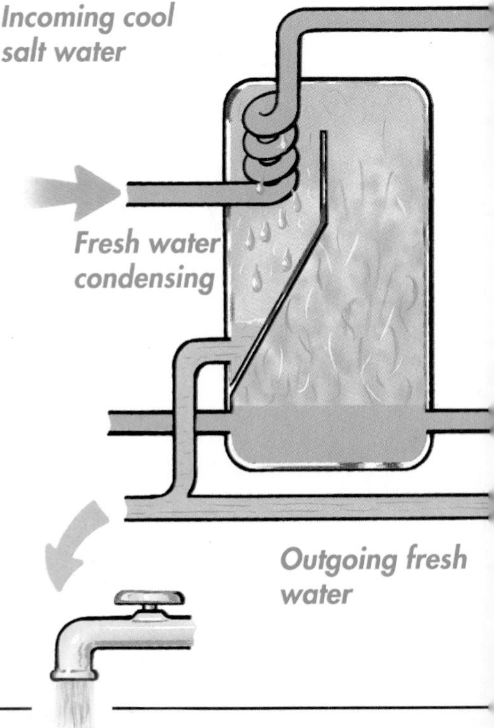

Focus on Technology

Making Fresh Water From Salt Water

Because of the need for more water, many people believe that we will have to depend on ocean water to increase our supply of drinking water. **Desalination** is the removal of salt from ocean water to get fresh water. Obtaining fresh water from the oceans is not a new idea. Aristotle described a method of evaporating sea water that Greek sailors used in the 4th century B.C.

The most common way of removing the salt from ocean water is a process called **distillation.** In distillation, ocean water is heated, causing the fresh water to evaporate. The salt is left behind. The water vapor is collected in a cooling tank, where it condenses back to water. This water is fresh water.

Distillation facilities are located all over the world. Coastal cities in North America, South America, Greenland, Europe, Africa, Asia, and Australia benefit from this process of producing fresh water.

Desalination plants can help reduce water shortages along the coasts. A large desalination plant, like the one at Ash Scuwaykh in Kuwait, can produce 284 million gallons of fresh water each day.

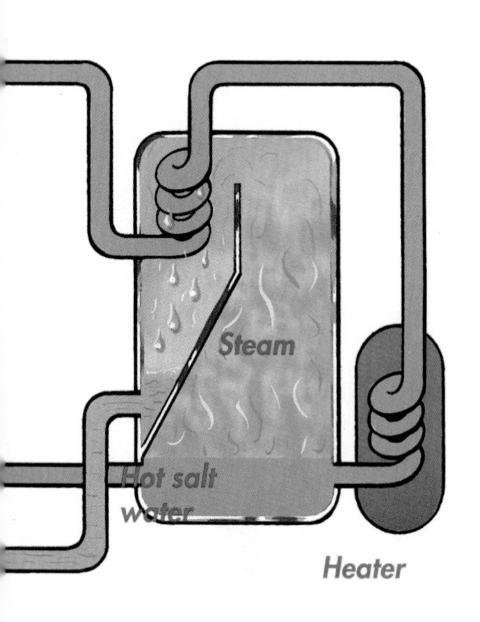

Salt Water and Plants

How does salt water affect plants?

What You Need

2 9-oz. cups, small spoon, salt, 2 plants, marking pen, *Activity Log* page 18

Mark one cup with an S and put one rounded spoonful of salt in it. Fill the cup with water and stir until the salt dissolves. Fill the second cup with fresh water. Put a plant in each cup and observe for three days. What effect did the salt water have on the plant?

The oceans are an important part of our planet. We depend on the oceans for food, energy, and transportation.

Shipping is still the most important method of transporting large loads of materials from one continent to another.

In the late 1980s, 20 percent of the world's oil was found in offshore wells beneath the ocean floor.

Have you ever eaten tuna, shrimp, or lobster? If you have, you have used the ocean for food.

Focus on Environment

Dolphin-Safe Tuna

Dolphins are mammals that often swim in the ocean with tuna. Fishing fleets search for dolphins with helicopters and speedboats, hoping to locate schools of yellowfin tuna. When they find dolphins, the fishermen throw out their nets expecting to catch tuna. But they not only catch tuna, they also catch dolphins. Since dolphins are air-breathing animals, they drown when trapped underwater in these nets.

In the 1980s the fishing industry killed 100,000 dolphins a year. Most U.S. tuna boats have stopped using methods that kill dolphins, but some still continue to use the nets. Since 1990, the largest U.S. tuna companies have refused to purchase tuna caught using methods harmful to dolphins. You can, too! Look for dolphin-safe labels on tuna that you buy at the grocery store.

Life Found in Coral Reefs

In addition to using the oceans for transportation, energy, and food, we can use oceans for recreation. Coral reefs are beautiful underwater gardens. Divers can enjoy the many life-forms that are found in the warm waters near Australia, Barbados, and Bermuda. Even people who don't dive can discover the beauty of the reefs by viewing the pictures that divers take or by taking a glass-bottomed boat ride.

Literature Link *Down Under Down Under*

Enjoy the memories of a fabulous diving adventure through the eyes of a 12-year-old. Select an ocean organism from this book that you would like to be. Write a short story about what it would be like for this plant or animal to live in its ocean environment.

Sponges are filter feeders. They get food by straining bacteria and other small organisms from the water.

Coral is formed by millions of tiny animals. Different types of coral make up the coral reefs.

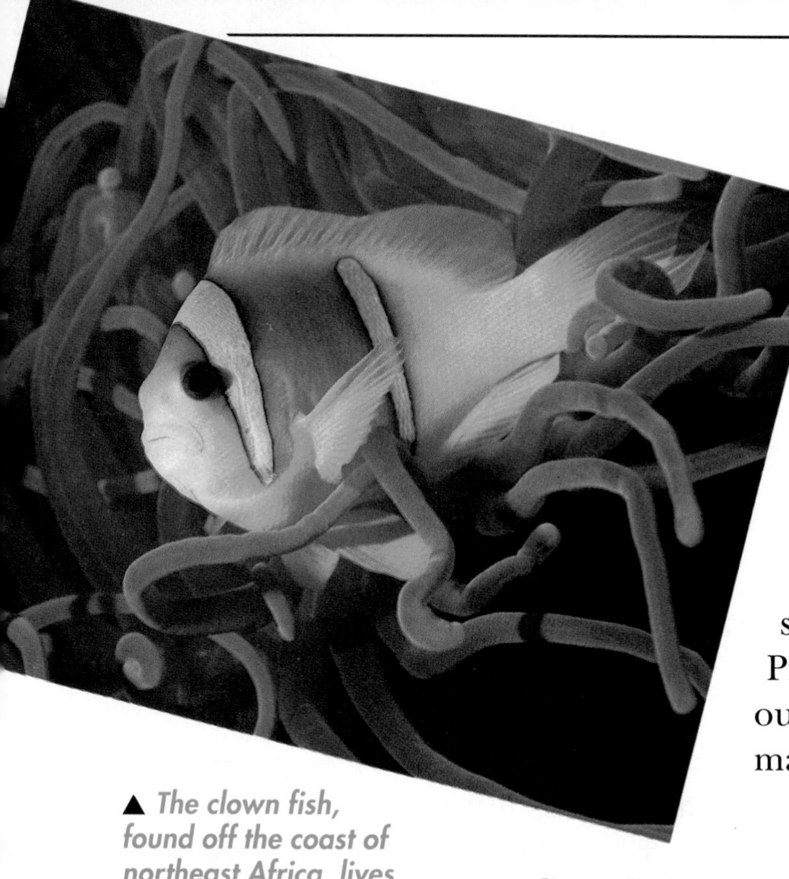

▲ *The clown fish, found off the coast of northeast Africa, lives peacefully within the sea anemone's poisonous tentacles.*

Language Arts Link

A Shorter Way To Say It!

Acronyms are words formed from the first letters or parts of a series of words. For example, NASA stands for National Aeronautics and Space Administration. ZIP stands for Zone Improvement Program. Use a dictionary to find out what SCUBA stands for. Now make up your own acronym.

Sum It Up

We interact with the ocean every day, from the foods we eat to the materials that are traded across the oceans. We rely on the fresh water, minerals, and energy sources that we get from the ocean. As long as we take care of the oceans, we can benefit from all they have to offer us.

▲ *Moray eels hide in rocks and coral.*

Critical Thinking

1. Describe the importance of water in your life.

2. What would Earth be like without the oceans?

3. You're planning to sail around the world in a small boat. How will you get drinking water when your supply runs out?

Protecting the Ocean

The more we explore the ocean, the more we realize the need to protect this valuable resource. For example, a scientist visited a faraway island in the South Pacific. Though no one lived on the island, garbage covered the beaches. The movement of the ocean carried the trash to the island.

Seals and otters are examples of animals that can be harmed by trash in the ocean. This seal has a fishing net caught around its neck.

Plastic items have been seen floating in the middle of the Atlantic Ocean. Animals sometimes get plastic pieces wedged in their stomachs or plastic rings tightened around their mouths or necks.

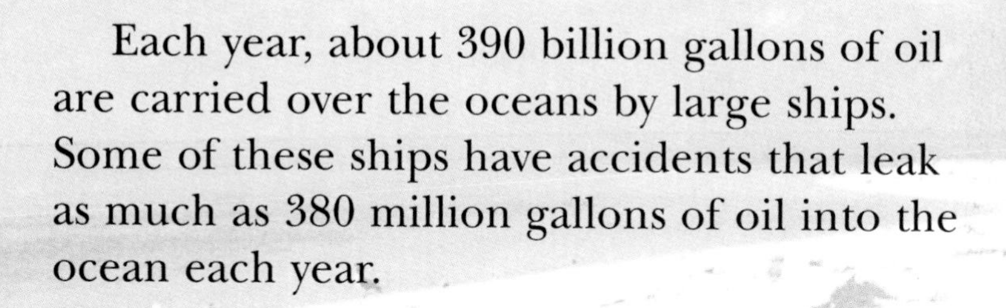

Each year, about 390 billion gallons of oil are carried over the oceans by large ships. Some of these ships have accidents that leak as much as 380 million gallons of oil into the ocean each year.

Oil spills can cause damage to water, wildlife, and the environment. Sea otters and birds that live near the shore are poisoned when they eat fish that have been contaminated by the oil.

With the amazing amounts of oil entering our oceans each year, it is important that we be able to clean up the oil spills. Unfortunately, it is difficult to clean them up. The oil is spread quickly by currents, wind, and ocean waves.

Chemicals are sometimes used to clean up oil spills. This method will help save wildlife on the coasts and beaches, but the chemicals may poison ocean life.

Focus on Technology

Oil: A Difficult Spill To Clean Up

A new method is being tried to clean up the oil spills in two recent oil spill disaster areas. During the summer of 1989, some beaches in Prince William Sound were treated with a bacteria designed to eat oil. In the summer of 1990, the bacteria was used to speed up the decomposition of oil before it could reach the Galveston coast.

Do the Try This Activity on this page to try to clean up an oil spill.

Tanker spill surrounded by containment device.

TRY THIS Activity!

Clean Up an Oil Spill

What You Need

sand, colored water, vegetable oil, small jar with lid, *Activity Log* page 20

Place sand, colored water, and vegetable oil in the small jar, replace the lid, and shake. Then plan a way to get the oil out of the system. You can use only materials that are safe—not harmful to you or the environment. Carry out the plan. Did it work? Would your plan work on a real oil spill?

People now realize that the ocean cannot continue to be a dump for trash, oil, and chemicals. Besides studying ways to clean up oil spills, people have looked for other ways to take care of our ocean and the organisms that live in it. Since 1988, an international treaty has banned the dumping of plastics from ships. Unfortunately, not all ships observe this new rule.

The International Whaling Commission has stopped all commercial killing of whales to save the population of whales from becoming extinct. The larger U.S. tuna companies are using different nets so that dolphins are not needlessly killed. Perhaps one way the number of oil spills can be reduced is by using radar and tugboats to guide the oil tankers through rough areas. Preventing oil spills would decrease the amount of oil that enters the ocean and decrease the harm done to ocean life.

Minds On! If you were a young environmentalist working to preserve ocean life from oil spills, what laws would you recommend to help prevent spills? ●

The ocean supplies the world with oxygen, water, food, and minerals. It also gives us the joy of knowing that animals like sea anemones and dolphins exist and the excitement of discovering new life in the ocean. If we want to continue to enjoy using the oceans, we must protect them and everything in them.

GLOSSARY

Use the pronunciation key below to help you decode, or read, the pronunciations.

Pronunciation Key

a	**a**t, b**a**d
ā	**a**pe, p**ai**n, d**ay**, br**ea**k
ä	f**a**ther, c**a**r, h**ea**rt
âr	c**are**, p**air**, b**ear**, th**eir**, wh**ere**
e	**e**nd, p**e**t, s**ai**d, h**ea**ven, fr**ie**nd
ē	**e**qual, m**e**, f**ee**t, t**ea**m, p**ie**ce, k**ey**
i	**i**t, b**i**g, **E**nglish, h**y**mn
ī	**i**ce, f**i**ne, l**ie,** m**y**
îr	**ear**, d**eer**, h**ere**, p**ier**ce
o	**o**dd, h**o**t, w**a**tch
ō	**o**ld, **oa**t, t**oe**, l**ow**
ô	c**o**ffee, **a**ll, t**au**ght, l**aw**, f**ou**ght
ôr	**or**der, f**or**k, h**or**se, st**or**y, p**our**
oi	**oi**l, t**oy**
ou	**ou**t, n**ow**
u	**u**p, m**u**d, l**o**ve, d**ou**ble
ū	**u**se, m**u**le, c**ue**, f**eu**d, f**ew**
ü	r**u**le, tr**ue**, f**oo**d
u̇	p**u**t, w**oo**d, sh**ou**ld
ûr	b**ur**n, h**urr**y, t**er**m, b**ir**d, w**or**d,c**our**age
ə	**a**bout, tak**e**n, penc**i**l, lem**o**n, circ**u**s
b	**b**at, a**b**ove, jo**b**
ch	**ch**in, su**ch**, ma**tch**
d	**d**ear, so**d**a, ba**d**
f	**f**ive, de**f**end, lea**f**, o**ff**, cou**gh**, ele**ph**ant
g	**g**ame, a**g**o, fo**g**, e**gg**
h	**h**at, a**h**ead
hw	**wh**ite, **wh**ether, **wh**ich
j	**j**oke, en**j**oy, **g**em, pa**g**e, e**dge**
k	**k**ite, ba**k**ery, see**k**, ta**ck**, **c**at
l	**l**id, sai**l**or, fee**l**, ba**ll**, a**ll**ow
m	**m**an, fa**m**ily, drea**m**
n	**n**ot, fi**n**al, pa**n**, k**n**ife
ng	lo**ng**, si**ng**er, pi**n**k
p	**p**ail, re**p**air, soa**p**, ha**pp**y
r	**r**ide, pa**r**ent, wea**r**, mo**r**e, ma**rr**y
s	**s**it, a**s**ide, pet**s**, **c**ent, pa**ss**
sh	**sh**oe, wa**sh**er, fi**sh** mi**ss**ion, na**ti**on
t	**t**ag, pre**t**end, fa**t**, bu**tt**on, dress**ed**
th	**th**in, pan**th**er, bo**th**
<u>th</u>	**th**is, mo**th**er, smoo**th**
v	**v**ery, fa**v**or, wa**v**e
w	**w**et, **w**eather, re**w**ard
y	**y**es, onion
z	**z**oo, la**z**y, ja**zz**, ro**s**e, dog**s**, hou**s**es
zh	vi**si**on, trea**s**ure, sei**z**ure

bathyscaphe (bath´ə skāf´) a small submersible vessel that carries a crew; used for undersea exploration

bioluminescence (bī´ō lü´mə nes´əns) light produced by a living organism

continental shelf (kon´tə nen´təl shelf) the broad edge of a continent that forms a shelf extending underwater from the shore to a depth of approximately 200 meters

continental slope (kon´tə nen´təl slōp) starts at the edge of the continental shelf and angles down to the deep ocean floor

current (kûr´ənt) concentrated flow of water due to temperature, salinity, or sediment changes

deep ocean current (dēp ō´shən kûr ənt) a deep water current that flows below surface currents

desalination (dē´sal ə nā´shən) the removal of salt from salty water to get fresh water

distillation (dis´tə lā´shən) to heat until evaporation takes place and then condense the vapor given off

gravity (grav´i tē) the pulling force of every object on all other objects

mid-ocean ridge (mid ō´shən rij) a chain of mountains that runs along the ocean floor

ocean trench (ō´shən trench) a long, deep, narrow canyon in the sea floor plain

ocean wave (ō´shən wāv) the up and down movement of surface water caused by wind

oceanographer (ō shə nog´rə fur) a scientist who studies the ocean

salinity (sə lin´i tē) the amount of dissolved salts in water

scientific method (sī ən tif´ik meth´əd) an organized plan of asking questions, gathering information, and finding answers

seafloor plain (sē flôr plān) large regions of the seafloor that are relatively flat

seamount (sē´mount) a large volcano rising above the seafloor

sonar (sō´när) sound navigation and ranging—an instrument used to detect underwater objects and to determine their location by means of sound waves reflected from or produced by the objects

submersible (səb mûr´sə bəl) a diving vehicle with an extremely strong body used for underwater research; some are designed to carry a crew, others are controlled by people on a ship at the surface

surface current (sûr´fis kûr´ənt) a current flowing at the water's surface, usually caused by winds

tide (tīd) the regular rise and fall of ocean water levels due to gravitational attraction of the moon

water cycle (wô´tər sī´kəl) the continuous movement of water (H_2O) in its different states through evaporation, condensation, and precipitation in the environment

INDEX

CREDITS

Photo Credits:

Cover, Animals Animals/Carl Roessler; **1,** ©Stephen Frink/The Waterhouse; **2, 3,** ©Comstock, Inc.; **3,** ©Platinum; **5,** ©Comstock, Inc., **6, 7,** The Image Bank/M. Friedel; **8,** (t) Earth Scenes/Sydney Thompson, (b) UNIPHOTO; **8, 9,** ©Michael Holford; **9,** Jack Stein Grove/Tom Stack & Associates; **12, 13,** ©Lynn Funkhouser/Peter Arnold, Inc., (insets) ©Studiohio; **14, 15,** David Young-Wolff/Photoedit; **16, 17,** ©Platinum; **18,** ©Comstock Inc.; **18, 19,** ©Comstock Inc.; **19,** (t) Anthony Howarth/INT'L STOCK, (b) ©Comstock Inc.; **20,** Animals Animals/Patti Murray; **21,** The Image Bank/Peter Frey; **22,** (t) ©Comstock Inc., (ml) Brian Parker/Tom Stack & Associates, (mr) Dave B. Fleetham/Tom Stack & Associates, (b) Joe McDonald/Tom Stack & Associates; **23,** (t) Animals Animals/Miriam Austerman, (b) Animals Animals/Zig Leszczynski; **24, 25,** UNIPHOTO; **26, 27,** ©Studiohio; **29,** ©Jim W. Grace/Photo Researchers; **30,** (m) ©Comstock Inc., (b) ©KS Studios/1991; **31,** ©Comstock Inc., **32,** The Image Bank/Terje Rakke, **34,** (t) Dave B. Fleetham/Tom Stack & Associates, (m) ©Animals Animals/E.R. Degginger; (b) F. Stuart Westmoreland/Tom Stack & Associates; **34, 35,** Animals Animals/Zig Leszczynski; **35,** Larry Lipski/Tom Stack & Associates; **36, 37,** ©Tim Rece/Animals Animals; **38, 39,** ©Platinum; **42,** (t) Culver Pictures Inc., (b) Photoedit; **43,** (t) ©Hank Morgan, Science Source/Photo Researchers, (b) ©Rick Friedman/Black Star; **44,** (tl) Animals Animals/Oxford Scientific Films, (bl) Norbert Wu/Tony Stone Worldwide/Chicago Ltd.; **45,** (t) Norbert Wu/Tony Stone Worldwide/Chicago Ltd., (b) ©Steinhart Aquarium/Photo Researchers; **46, 47,** ©John Lythgoe/Planet Earth Pictures; **48, 49,** ©Studiohio; **50,** ©Russ Kinne/Comstock Inc.; **52,** (t) David Lavin/UNIPHOTO, (m) ©Claymore Field/Planet Earth Pictures, (b) Jon Feingersh/ALLSTOCK; **53,** ©M. Webber/Planet Earth Pictures; **54,** Comstock Inc.; **55,** (t) ©Peter Scoones/Comstock Inc., (b) ©Franklin Viola/Comstock Inc.; **56,** ©Frans Lanting/Photo Researchers; **57,** (inset) Animals Animals/R. Ingo Riepl, ©Frans Lanting/Photo Researchers; **58,** The Stock Market/©Shorty Wilcox, 1990; **59,** ©Stephen Frink/The Waterhouse.

Illustration Credits:

6, 14, 24, 36, 46, Jim Theodore; **16, 26, 38, 48,** Bob Giuliani; **20, 28, 33, 40, 41,** John Edwards; **50, 51,** James Shough